# For The Love of Oil

## The Fleecing of the American Consumer by Big Oil Companies, Politicians, and Wallstreet Commodity Traders

### J.C. McElroy

Bloomington, IN  Milton Keynes, UK

*AuthorHouse™*
*1663 Liberty Drive, Suite 200*
*Bloomington, IN 47403*
*www.authorhouse.com*
*Phone: 1-800-839-8640*

*AuthorHouse™ UK Ltd.*
*500 Avebury Boulevard*
*Central Milton Keynes, MK9 2BE*
*www.authorhouse.co.uk*
*Phone: 08001974150*

*First published by AuthorHouse 6/20/2006*

*ISBN: 1-4259-4254-7 (sc)*

*Printed in the United States of America*
*Bloomington, Indiana*

*This book is printed on acid-free paper.*

# TABLE OF CONTENTS

**PRELUDE**

---

# PRELUDE TO CORRUPTION;
# THE POLITICS OF OIL

From 1995 to 2000 fourteen of the top fifteen oil companies merged in some fashion. Most notables were in 1999 when Exxon and Mobil merged. Both were originally Standard Oil companies. In 2000 Chevron acquired Texaco. Both were also formed from the breakup of Standard Oil. Also in 2000 British Petroleum (BP) merged with former Standard Oil companies Amoco, and Atlantic Richfield (ARCO).

All mergers and acquisitions sailed through congress. The FTC and the SEC requiring them to only sell off small portions of their companies refining capacity to other big oil companies not involved in the particular merger. It was nothing short of a shell game of oil refining ownership between the largest oil companies once it was all completed.

As some have stated, the oil and gas company mergers of the late 1990's simply put Standard Oil back together again. By allowing most of its former companies to merge into a handful of companies, they have united in effort to become a powerful entity that controls oil and gas prices by manipulating the refining and distribution process, much the same as Standard Oil once did. In one respect, they have created a corporate cartel, much more powerful than OPEC, but one that OPEC members benefit from.

In May 2001 Senator Carl Levin of Michigan pointed out in a letter to the General Accounting Office, now the Government Accountability Office, (GAO) that the oil and gas company mergers of the late 1990's had had an adverse effect for consumers. He expressed great concern that price manipulation was ongoing and had only been made easier by the mergers. The GAO in an earlier report had also concluded that oil and gas company mergers of the late 1990's had led to higher prices.

And yet politicians have done nothing about it. Unless you count the tens of millions of dollars they have taken from the oil companies to do nothing about it.


The goal of this book is to educate and explain in basic terms how the oil companies and others have driven up the cost of oil products and natural gas. It will also give you the information needed to help bring about political change.

## CHAPTER I

---

# A BRIEF HISTORY OF OIL

## STANDARD OIL

John D. Rockefeller co-founded Standard Oil in 1870 by uniting several of his oil businesses. In 1882 Standard Oil became Standard Oil Trust and by 1890 Standard Oil controlled over 90% of refined oil in the U.S. It made J.D. Rockefeller the richest man in the world.

In 1890 the U.S. Congress passed into law the Sherman Anti-Trust Act, in order to regulate and control large companies that had grown into monopolies. In 1892 the Ohio Supreme Court declared Standard Oil Trust a monopoly. Standard Oil fought the ruling and re-organized into a holding company in New Jersey, named Standard Oil of New Jersey.

Standard Oil and the other big monopolies were not brought down by the Sherman Act initially. Just like today, big business had bought its influence into both political parties, making it nearly impossible to fully regulate an industry in the best interest of the average citizen. Standard Oil was brought down by a constant media attack on business monopolies, its effects on the economy, and the corrupt politicians that allowed it to persist.

In 1894 Henry Lloyd wrote "Wealth Against Commonwealth". He attacked monopolies by using Standard Oil as the model of greed, political influence, and excess profits. This led others to write books and articles in the coming years about big business and political corruption.

The 1902-1912 period was know as the muckraking era for its constant media exposure of mega-business and corrupt politics. The best known of the 'muckrakers' was Ida Tarbell, who wrote several articles about Standard Oil and its business practices to control the price of oil and oil refined products. Others wrote on issues such as medicines, and the corruption in the U.S. Senate and House of Representatives.

The newsmedia exposed what most Americans had long known. That political greed and big business greed were hand-in-hand. The media, as always, held the power to dig for the truth and report it. They did with countless stories. The public became outraged and demanded political change and the breakup of the Standard Oil monopoly.

In May 1911, the U.S. Supreme Court ordered the breakup of Standard Oil into 34 separate companies, mostly devided by the various states they operated in.

We need the same effort by today's media. Unfortunately the relaxing of FCC laws in the 1990's has allowed for the buying, selling, and merging of news stations all the way down to the small town level by big business conglomerates. Now news is simply pre-packaged for sensationalism. Unfortunately the days of the 'gumshoe' reporter who went out and dug up and investigated corruption in order to report the truth seem to be gone. Today's media simply report what they are told by industry analysts who are little more than a go-between for the media and the business in question. The answers always sound more believable coming from an "Analyst" than a corporate executive. Standard Oil used similar tactics back in the day.

## U.S. & BRITISH INVOLVEMENT IN THE MID-EAST

In 1933 the U.S. government paid King Saud of Saudi Arabia for the rights to drill for oil. Many thought this to be a big waste of American tax dollars, as most geologist did not believe Saudi Arabia had oil.

In 1938 American's struck oil in the Saudi desert and the U.S. government gave big oil companies incentives to enter the middle-east and drill for oil. A 50/50 split on oil profits between the oil company and the country the oil field was in was the standard practice.

While the Americans were in Saudi Arabia the British were developing a vast oil industry in Iran. During the same time period Hitler began his conquest of neighboring European nations. Germany had no oil reserves, or major oil exploration companies to speak of, and thus had to buy it's supply.

The Nazi-Germany invasion of Europe would require vast amounts of fuel, and a steady supply of it. The German's turned to invention and innovation for a reliable supply of fuel. They began making fuel from processing coal refuse to supplement their supply, and did so until their defeat in 1945.

If to the victor's belong the spoils of war then German fuel technology was not something the U.S. and British governments had an interest in. They were already deeply invested and financially tied to the big oil companies drilling for oil in North Africa, also known as the Middle-East.

The British continued to control the oil reserves in Iran until 1951. During that year the Iranian government simply took over from the British by nationalizing the oil industry. This was the hazard of doing business in a country with an unstable political environment. This shift in control of the oil fields to the country of origin gave the ruling elites greater wealth and power to dictate terms. It also helped lead to the formation of OPEC in 1960.

By the time oil was discovered in Libya by U.S. companies the middle-east nations had become bolder and wiser in negotiations. Libya was the first country to dictate terms for drilling and pumping it's oil to foreign companies. Thus marked the end of the major oil companies control over oil in Africa. A tight grip they had held with the aid of the U.S. and British governments help.

After taking the lead in policing the mid-east for almost a century, the British government pulled out in 1971, leaving the Arabs to fight amongst themselves. The U.S. government moved swiftly to try and bring a stable influence to the region once the British withdrew. The U.S. backed the Shah of Iran in hopes of providing a western friendly ally to govern the country, and suppress the radical religious factions who were Anti-American.

## OPEC

The Organization of Petroleum Exporting Countries (OPEC) was founded in Baghdad, Iraq in 1960. Iran, Iraq, Kuwait, Saudi Arabia, and Venezuela were the founding members. By 1971 Libya, Qutar, Nigeria, Indonesia, and the United Arab Emeriates had joined the oil cartel.

As with any cartel, OPEC's intentions are to control the supply of their product, oil, from their member nations in order to keep up a steady price. OPEC did not gain any real acceptance as a marketing force until the mid-seventies oil embargo, but even that was not of their doing.

In 1980 OPEC switched to a quota system based on a member country's size of oil reserves. The larger the nation's reserves the more they would be allowed to pump. As a result by the mid 1980's all OPEC nations had greatly increased their stated oil reserves. Some of the increase was in fact due to new oil field discoveries, and some due to new technology that allows for a larger percentage of oil to be pumped from wells. It is impossible to know what the true oil reserves are since most OPEC nations do not allow for outside audits or examinations of their oil field data. But there are obvious profits to be made for each country in overstating it's oil reserves under this system. It should be noted that Iraq has not been apart of the quota system since 1998.

In more recent years OPEC has tried what they term a world average basket oil price in order to maintain a steady average price of crude oil at $22 to $28 a barrel. However, they suspended this method in 2005, as they had not been able to control oil prices which now have nothing to do with the supply they partially control, and everthing to do with the refining of oil which is all done in the U.S. by American and British companies.

It is easy to see why cartels are illegal in the U.S. It goes against our free market system, and falsely drives up the price of product. Just think how high oil and gas would go if major oil companies were allowed to conspire to control the amount of oil that is refined into usable products. But they have been allowed to do this and it is a large part of the equation for current high energy prices.

OPEC has never been fully effective in controlling oil output and world prices. This is mainly due to Non-OPEC oil producing countries producing more oil and OPEC's own Saudi Arabia producing more than the stated limits everytime the U.S. needs more oil.

## ARAB OIL EMBARGO

In the fall of 1973 Israel was attacked by Syria and Eygpt and started what was known as the Yom Kipper War. The U.S. and most European nations supported Israel and as a result oil was used as an economic weapon by the Arabs against the U.S. in retaliation. The oil embargo caused a supply panic in the U.S. and oil prices soared from $3 to $11 a barrel.

The embargo lasted until March, 1974, and then oil prices quickly fell. There was no actual oil supply shortage, only an embargo for oil coming from certain mid-east nations to the U.S. Many blamed the big oil companies for holding oil supplies already stockpiled and making huge profits from the embargo, which they did.

The oil embargo led to a more stable practice of drilling for oil and natural gas outside of the middle-east by all companies. With new technological developments the European oil companies were soon producing more oil from the North Sea than were from many of the OPEC nations. The embargo also pushed congress to pass the Alaskan Pipeline Bill to open up oil fields in Alaska. The pipeline began moving oil in 1977.

By the early 1980's 25% of the U.S. oil supply was coming from the Alaskan pipeline. There was plenty of oil, just as there always had been. But now from a reliable domestic source. In response OPEC installed production maximums to help hold up oil prices. It did not work. With American's driving more energy efficient cars and small trucks, true market forces took over and oil prices fell from $29 to $10 a barrel in 1986.

## IRAN AND IRAQ

In 1979 the U.S. backed Shah of Iran fell from power and was forced into exile by the Iranian Revolution. The Anti-American religious leader Ayatollah Khomeni took over as the leader of Iran, and caused amongst other things a second oil price shock in less than ten years.

Once again there was no actual shortage of oil, but prices began to rise. With the Shah in exile, the U.S. Government which protected oil company interests had no ally, or puppet leader, in the region. This gave Iraqi dictator Saddam Hussein the opportunity to invade his Iranian neighbors to try and seize their oil fields. After a decade of fighting, and an estimated one million killed, Iraq withdrew from Iran, ending the Iran-Iraq war in a stalemate. As a result of the war with Iraq, and the change in Iranian leadership, Iran's oil output is still only two-thirds of what it was when the Shah was in power.

## THE GULF WAR

Saddam Hussein borrowed heavily from his middle-eastern neighbors in his failed quest to take over Iran's oil fields. Iraq owed it's neighbors an estimated $100 billion, and Saddam asked for debt forgiveness. Some leaders were willing to work with the Iraqi dictator, mostly out of fear. But the tiny country of Kuwait refused.

In 1991 Saddam Hussein's army invaded Kuwait with the same intent he had invaded Iran. To capture it's vast oil fields or destroy them if he could not take control of the oil.

Saddam's army was quickly defeated and driven from Kuwait via a U.S. led United Nations mandate. It is important to note that although U.S. forces did most all the heavy fighting, it came under a U.N. coalition that had the objective to force Saddam's military from Kuwait and to restore the oil fields there. Not to overthrow Saddam Hussein as the leader of Iraq, as U.S. leaders had pushed for.

Oil prices spiked only briefly during this event, as the Iraqi army surrendered quickly and U.S. and British companies rushed in to put out the oil field fires and restored production.

## THE OVERTHROW OF SADDAM HUSSEIN

Many claimed the invasion of Iraq in 2003 by U.S. forces to overthrow Saddam Hussein and liberate the Iraqi people was really an invasion to take over Iraqi's oil fields. As much as we would like to believe this is not true, it has some truth to it.

The claim of a U.S. invasion to take control of the vast oil reserves sounds similar to what Saddam Hussein did to Iran in 1980 and to Kuwait in 1991. Whether we went to Iraq for control of it's oil or to overthrow a dictator that voiced his hatred for America, the secondary result is the Iraqi people have an opportunity to be a free nation. However, what is trying to be kept out of the mainstream media is all the contracts given to Vice-President Dick Cheney's former employer, Halliburton.

Halliburton and it's many subsidairy companies have gotten the contracts to rebuild the oil field infastructure in Iraq, as well as receiving the vast majority of contracts to feed and house the troops. Unfortunately, the longer the conflict goes on in Iraq and elsewhere in the middle-east, the more billions of dollars Halliburton and it's subsidairies will make from the U.S. taxpayers. Without a doubt the war is a financial windfall for Halliburton, and it will be discussed in further detail in chapters three and four.

The facts show that the U.S. Government had a decisive plan to take control and rebuild the oil fields, but had little or no advanced planning to take control of the cities of Iraq and bring peace and security there.

It is important to point out that the U.S. got no oil from Iraq prior to the invasion, and as of this writing, still gets none of the oil from Iraqi oil fields. Iraqi oil is sold to the countries it had agreements with prior to the invasion, namely neighbors Turkey and Jordan. European countries France and Germany that Iraq dealt with under the scandal riden U.N. food for oil program also receive it's oil, which will be discussed in chapter three.

In short, the invasion of Iraq was to overthrow Saddam Hussein, Liberate it's people, take control of it's only valuable resource, oil, and make billions for well connected companies like Halliburton. But maybe not in that order.

One thing is clear. It was much easier to take control of the oil fields than to bring peace and security to the streets of Iraq. That alone should be a good indication that the Iraqi people want U.S. forces there to allow for freedom. If not, surely they would have put up a fight at the entrance to the oil fields. Their only national source of income.

## NATURAL GAS - A BRIEF HISTORY

Natural gas, just like oil, has never really been in short supply at the source. It's supposed shortage is also due to gas companies manipulation of the supply chain.

Natural gas, unlike oil, can be created or produced from many other sources other than the naturally occuring form we extract from the earth. Some of the sources are coal, plant matter, manure, and landfills, from which it can and is being produced.

The British were the first to use man-made natural gas. In the late 1700's they produced gas from coal and used it to light houses and in streetlamps. Naturally occuring gas was first discovered in America in the 1620's by French explorers who saw Indians igniting it as it came from the ground. In the 1800's gas wells were dug in the American NorthEast to use gas. The first American natural gas pipeline was built in the late 1890's from North Central Indiana gas wells to the city of Chicago. But it took until the 1940's before there were reliable pipeline building materials and technology to provide a secure transportation of the gas from the wellhead to major cities.

Congress first began to regulate the natural gas industry in 1938. It had become a common fear that this growing source of fuel would become a monopoly, just as Standard Oil had done with oil. This rational fear led to the Natural Gas Act, which regulated a restriction on the price charged to customers. In the early 1990's the regulations were relaxed and the natural gas industry went through a period of deregulation and expansion.

With new developments in technology, vast amounts of gas have been discovered since the 1980's, as well as the ability to tap into gas fields once deemed unrecoverable. The U.S. Energy Information Administration and naturalgas.org., a natural gas information supply association, state that the common misconception is that we are running out of natural gas. In December, 2005, the U.S. Government reports showed natural gas stockpile levels in the U.S. were at five year highs. But so were prices. The Natural Petroleum Council's estimates that there is enough natural gas in the U.S. to meet our growing energy needs for at least 75 years. Yet oil and gas industry executives state that our domestic supply is running out. This is not totally true. Of course we are continuing to use up what we burn, just as they continue to find new deposits of gas and ways to produce it from other sources. The sky is not falling, as gas and oil industry executives would like us to believe. They make their statements of half-truths for two reasons. To justify what are supply manipulated prices, and to try and gain support for the right to drill for gas and oil on government restricted areas. Most notably in Alaska and in the Rocky Mountains.

Natural gas and propane, a by-product of the extraction of natural gas, are just as advertised. A clean burning reliable source of energy. Natural gas is not only used to cook with and warm homes, but it is also used to generate electricity at power plants, make agricultural fertilizers, and is also used in manufacturing plastics.

The United States is the number one consumer of natural gas, just as it is with oil, using 25% of the worlds annual output. However, unlike oil, the U.S. produces almost 85% of the natural gas it consumes. New reserves of natural gas are constantly being discovered all over the world, as technology continues to make advancements. The U.S. is the number one energy consumer in the world, but it also ranks third in world oil production, behind only Saudi Arabia and Russia.

# CHAPTER 2

---

# THE CHINA FACTOR - ALL HYPED UP

## COMMUNIST COUNTRY AND CHEAP LABOR

If the oil industry analysts aren't blaming hurricanes or the threat of terrorist for high energy prices then they blame China. In the past decade many American manufacturing jobs have moved to places like Mexico or China. But lets not forget why. American workers are paid in dollars-per-hour and Chinese workers are paid in cents-per-hour. It is that simple.

China is still a communist country, not a democracy. It's citizens have few civil rights. Almost half of the businesses are state owned entities (S.O.E.'s), that operate at a loss. More importantly, for a country to be a world economic powerhouse it's citizens have to have discretionary income to spend. The ability to buy goods and services as well as such tangible items Americans often take for granted such as cars, homes, and appliances. The average Chinese factory worker makes less than fifty cents per hour in U.S. dollars, and has no discretionary income. They can't even afford to buy the cheap goods or electronics they make. Only the government officials and a few business elites have the income to afford the single family apartments and other tangible items we call necessities.

## CHINA NO JAPAN IN THE MAKING

China is a vast country with a great history and culture. It has hard working people who understand sacrifice and loyalty. But it's economy in no way resembles Japan's. Many Wallstreet stock traders would like to convience everyone that the Chinese economy is a sure bet and the next Japan in the Asian economy. But nothing is further from the truth.

After we defeated Japan in World War II we rebuilt their country in our own image. A democracy with capitalistic values. That is why the Japanese economy is second only in the world to the U.S.

No matter how often they show the handful of streets in Shanghai with the big buildings and neon lights, that is not China as a whole country. The truth is the majority of China's 1.2 billion people live in the country, far from the cities. They eek out a living, more of a survival, from the land. This has been the tradition for thousands of years.

True, China's rural youth are moving to the cities in record numbers looking for education and any work available. But the fact remains that China's per capita GDP does not even rank it in the top 100 countries in the world. In otherwords, China is a third world country, and will remain a third world country as long as it's communist leaders allow it's workers to be paid such low wages that do not allow for discretionary spending. Remember, we never feared the communist Soviet Union taking over the world economically.

What China has to offer is cheap labor, few environmental laws, and no labor or class-action lawsuits. The Chinese government determined years ago that it needed foreign (U.S.) investment in China in order to have a sustainable economy. In short, China needs the U.S. business investments far more than the U.S. needs anything China produces. Cheap labor can be found in many places in the world. The Chinese economy would dry up overnight without American investments.

In a communist country the government is the economy. When one goes broke the other falls like a house of cards. Consider the former Soviet Union as an example. The difference with China is it has allowed for foreign investment where other communist countries remain isolationist in order to better control their people.

## SO WHERE'S THE GROWTH COMING FROM?

China pushed for a most favored nations trading status from the U.S. Government for years. In the late 1990's the Clinton Administration signed a trade agreement with the Chinese government. Thus gave a greater opportunity for U.S. companies to locate in China and take advantage of their cheap labor force, no E.P.A. regulations, or any of the other countless costs of doing business in the U.S. This opened the door for companies like Wal-Mart and the companies that supply their products to re-locate their manufacturing to China. Remember the brief scandal of Al Gore taking campaign contributions from the Chinese government for the 1996 Clinton-Gore re-election. Now you know why the Chinese government went to great links to make a campaign donation.

The move to China has been a boom for U.S. retailers doing their manufacturing there, as well as the Chinese government. In response, the government has built more factories and infastructure in the eastern coastal region of the country, where most all manufacturing and businessess are located.

The Chinese government continues to spend a great deal of money to support the many S.O.E.'s, and on energy to run the factories. Many independant economists and outside investors are skeptical about China's continued economic growth in relation to previous years. It is always easier to put up big numbers of percentage growth when you start at a number near zero.

## CHINA'S ENERGY RESOURCES

Until a few years ago China prided itself as being non-dependent of foreign energy sources. China's main source of energy is coal, which accounts for 65% of it's energy needs, and the coal mining industry is one of the nations largest employers. China is the worlds largest coal consumer, as most all of it's factoriers and powerplants are coal fired. This, along with it's crowded cities that lack sufficient sewage and trash disposal, is the reason why China has 7 of the 10 most polluted cities in the world.

Recently, the Chinese government has shown a strong interest in coal liquification technology. The ability to convert coal and coal refuse into liquid fuels, an old process now gaining new worldwide interest as oil and gas becomes more expensive.

Natural gas consumption accounts for only 3% of China's total energy usage. China does have vast natural gas reserves in the northwest region of the country. However, it would take a huge investment in infastructure in both roads and pipeline to get the gas to the eastern coastal areas where the cities and factories are located.

China has three major oil and gas companies which are majority owned and controlled by the government. Through it's oil companies, China has agreements with countries such as Iran and Sudan that won't sell oil to the U.S. The only major middle-eastern oil country that sells to both the U.S. and China is Saudi Arabia. In 2005 China reportedly bought 17% of Saudi Arabia's oil supply that was not designated for the U.S.

As with any communist country money is a major hurdle to overcome. In order to bring in much needed foreign capital for oil and gas development, the Chinese government has allowed all three oil companies to have partial stock I.P.O.'s in the recent years prior to 2005. The major western oil companies that control the U.S. and European oil and gas markets bought the majority of shares offered. Thus giving them a large interest in the oil and gas development for the Chinese market.

China is the largest country in the world, but accounts for just 10% of the world energy consumption. China ranks a distant second to the U.S. in total world oil consumption at just 7%, compared to 25% for the U.S. It should be noted that China is a vast country with a great amount of untapped oil, natural gas, and coal due to their lack of money, technology, and infastructure.

## CHAPTER 3

---

# POLITICS FOR SALE BY THE DEMOCRATS AND REPUBLICANS

## SPECIAL INTEREST MONEY IN POLITICS

The result of special interest money in politics is corruption and higher prices to consumers via unfair business practices. This is always the goal and result of any industry or business that lobbies the political machine known as the Democratic and Republican parties which make the laws. Nothing shows this better than the major oil and gas companies being allowed to merge. The result is higher energy prices based on their false claims of oil shortages for various reasons. Government agencies reports show just the opposite. Record oil and gas stockpiles. The only shortage is in gasoline which is in a shorter supply relative to oil supplies due to a collective effort by oil companies to reduce refining and transportation to inflate prices. Similar reasons as to why Standard Oil was finally broken up in 1911.

Unfortunately special interest money in politics is nothing new. It is nearly impossible to keep this kind of money out of politics since the politicians and political parties have grown accustomed to it.

There will always be elected officials willing to take what amounts to bribes. Usually in exchange for a vote, a government contract, or the relaxing of the very laws set forth to protect the average citizen from price fixing by the large industries. Once uncovered by the media the scandal unfolds.

The broader question is what happened in the 1990's that allowed for the current scandals? The answer is simple, yet complex. With the 1990's came an era of relaxed government policy changes perpetrated by special interest money via the change in campaign finance laws. Remember the McCain-Feingold Bill to limit these 'contributions'. It was dead before the ink dried. The political parties made sure of it. Pulling the strings of their respective politicians to ensure themselves a way around it should it become law. The same politicians you continually elect and pay to protect your best interests.

Through what some would call a backdoor method, the Republican and Democratic parties have created what is known as 527 organizations. These money gathering entities have allowed the parties to open their doors to accept even larger donations than ever before from big industry influence. With all this money flowing into political parties and then down to the candidates they want to win, who could lose? The U.S. citizen as always.

Lobbyists working for large corporations are paid to 'influence' politicians and therefore politics itself for greater industry profits. They are thriving under the current system, and have always done well.

The pharmaceutical industry wants to be able to continue to charge the American consumer three times the price for the same drug sold in Canada and elsewhere in the world. They also want to keep you from bringing medications in from Canada as well. One has to ask, if NAFTA is suppose to be so great for Americans then why doesn't it include prescription drugs.

In relation to the subject matter of this book, the big oil and gas companies want to be able to charge a bigger price for their energy products. They do it by manipulating the refining of oil and distribution of natural gas. They have shut down refineries and operate the rest at far less than 100% capacity. And they are doing it with government officials knowledge. Our government has agencies that track and report on all energy sectors. The U.S. Energy Information Administration has reported the refining capacity shortages and other discrepancies the oil companies report to the media. But it seems to fall on deaf ears.

## GOVERNMENT AGENCIES HARD AT WORK FOR WHO?

The Federal Trade Commission (FTC), Security and Exchange Commission (SEC), and the Federal Communications Commission (FCC), are suppose to regulate large businesses and industry, stocks and bonds, and the media in this country for the benefit of fair business trade and the benefit and protection of the American citizen. During the 1990's changes were made to the laws that guide these agencies. They have thus aided in the energy scandal, as well as other corporate wrong doings that came about as a result of a decade of merger-mania they gave credence to.

The relaxing of the FCC rules regarding how many t.v., radio, and newpapers a person or corporation can own is a factor in the overall manipulation. The controlling of the media by a handful of conglomerates is not a scandal. However, it has proven to limit the once independent opinion and stifle the investigative reporting that use to be the backbone of the newsmedia. Today news is quickly and neatly packaged in a cookie cutter approach to reporting that gives more attention to sound bites than exploring the facts.

First Amendment rights have been quietly stepped on for the interest of media conglomerates who decide what is newsworthy. Not necessarily what is in the best interest of the country or what citizens should know. Unfortunately, today's newsmedia is more interested in investigating and reporting what might be scandalous or newsworthy based on its level of dirty laundry or tradegy. In otherwords, it is easier to cover the trials and tribulations of Michael Jackson or Mike Tyson than to investigate the true cause behind higher energy costs. They simply relay to the public what the big oil companies tell them, instead of digging for and exposing the truth in order to bring about justice.

In the 1990's the FTC and the SEC allowed the mega-mergers with practically no oversight, and certainly no foresight. In regards to the oil and gas industry, they allowed the big mergers of the Standard Oil offspring companies by making small stipulations in refining capacity that were meaningless. They required the merging companies to sell off small portions of their oil refining facilities to other big corporations not involved in the particular merger, thus cutting out the small refining companies even more. The end result was simply a trading of some refining facilities between the top 15 oil companies that merged into 5 huge companies from the mid-1990's through 2000. Smaller oil refineries gained no capacity, as was the intention of the oil giants. Thus, true market competition has been squeezed out. Exactly what the big oil company executives had set forth to do by merging with favorable government stipulations. Once again, the oil lobbyists got for the industry what they were paid to do.

## PUBLIC CITIZEN GOES TO WASHINGTON

Many public interest groups and watchdog organizations continually collect factual information and congressional voting records to remind members of congress who they are suppose to represent. These groups lobby the media with facts on behalf of the American taxpayer in order to try and keep elected officials honest.

The corporate industry lobbyists use money, expensive vacations, and in some cases kickback schemes to garner votes and political influence. Guess who is winning this battle.

In September 2005, Tyson Slocum, acting director of the non-profit public interest group Public Citizen's critical energy mass program, testified before the U.S. Senate concerning high gas prices. He reminded the Senators of the following facts:

* In 1999 U.S. oil companies made 22.8 cents per gallon net profit. In 2004 they made 40.8 cents per gallon net profit.

* The oil industry gave $52 million dollars in campaign contributions from 2001-2004. In return the Senate voted 74 to 26 to pass the 2005 Highway Bill which gave the oil companies $6 billion dollars in taxpayer funded subsidies and tax breaks.

* The U.S. is in fact the 3rd largest oil producer in the world.

* OPEC nations supply just 14% of U.S. oil needs.

* The Senate voted down higher U.S. gas mileage standards for automobiles that would have cut fuel consumption by one-third in 10 years.

Mr. Slocum also made the following recommendations to rectify the current energy situation. Many of these recommendations have been made by others as well.

* Impose a windfall profits tax on oil companies.

* Investigate all oil companies and their executives.

* Re-evaluate all oil company mergers of the past 10 years.
* Impose higher fuel economy standards for all automobiles.


Without question these are all things that the U.S. Congress should do and should have done years ago, and didn't. Here's why.

Nothing seems easier and makes more sense than to simply increase fuel economy on cars and trucks. Once again the U.S. auto industry lobby does not want it. They say it costs too much money to make the changes. But the fact is for the better part of a decade you can buy a programmer, or 'chip' as they are sometimes refered to, in the aftermarket sector for $300 to $500 dollars. They increase your engine horsepower and at the same time give you 2 to 5 more miles per gallon by making the engine run more efficiently. The programmers are mainly sold for diesel pickups, but gas engines are fuel injected as well, and are programmable. This technology is not new or complicated, and the average person can buy and use the programmer. In short, if it is being done in the aftermarket sector then it can be done at the factory for even less per vehicle.

Keep in mind the thinking of U.S. automakers. They still advertise how fast they can get you from 0 to 60. But they don't offer to pay your ticket for reckless driving. The Japanese car makers build and advertise based on reliability, fuel economy, and luxury at an affordable price. And you never see them have to offer huge discounts to sell their cars.

Imposing a windfall profits tax would be logical as well. Although impossible to get the congressional votes with all the money the oil industry has poured into both political parties. A full investigation of oil companies and their executives by the Feds would quickly lead back to members of congress. So don't look for that to happen anytime soon. Regulating energy commodities and oil company mergers is a must. It will be discussed in full in chapter four.

# THE UNITED NATIONS FOOD FOR OIL SCANDAL

The U.N. sponsored oil for food program began in 1996 to help the Iraqi people gain the much needed food their country could not produce or purchase. Under post Gulf War U.N. sanctions, Iraq was limited on the amount of oil it could sell. The oil for food program allowed Iraq to sell oil under U.N. guidelines to finance the purchase of humanitarian goods, such as food and medicines.

The program turned out to be a cash cow for the very man the sanctions were imposed upon. Saddam Hussein made billions under the table along with the countries he dealt with via this program. These countries were the same countries that were vocally against further U.N. sanctions, and the U.S. invasion of Iraq in 2003. Namely France, Germany, and Turkey.

No one lost their position at the U.N. over this scandal. Even though the U.N. Secretary-General, Kofi Annan's son was directly involved in one of the top companies set up to take advantage and profit from manipulating this program.

This corrupt program ended and was exposed with the invasion of Iraq by U.S. forces. Some might say it ended one scandalous program at the U.N. only to make way for many more involving U.S. oil companies and their subsidairies getting no-bid contracts in Iraq.

The idea of trading oil for food is noble and justified. It is a true shame that people are not held accountable for their actions at an organization that is suppose to stand for world peace and integrity. Who does the U.N. think they are? The U.S. Congress!

# REVOLVING DOOR APPOINTMENTS

Even when independent of one another, political lobbyists and revolving door appointments are highly destructive to our democratic process and free market economy. Unfortunately in the last decade they have by their own nature become united.

Revolving door appointments come in one of three ways. Government job to industry position. Government position to lobbyist. Or industry position to political appointment.

The appointment of corporate executives or lobbyists to public or political jobs, and the movement of government officials to high paying private sector corporate jobs with companies doing business with the government will never lead to a positive outcome for the taxpayer of this country.

In the aftermath of hurricane Katrina one of these high-level appointments was exposed. Michael Brown, who had worked in the horse racing industry, was head of FEMA. With no prior experience to qualify him for the position, he was simply appointed based on political favors. One would think it should be a requirement to have experience in dealing with large scale disasters to hold such an important position. But these type of appointments have gone on for decades.

Most revolving door appointments are never as visable. Riley Bechtel, CEO of Bechtel Group, Inc., was appointed to President Bush's export council. Bechtel does contracting work for the government in the areas of water, nuclear energy, and public works projects. Former Secretary of State George Schultz is a board member of Bechtel Group, which has garnered many lucrative government contracts since 2001.

The best known revolving door appointment is Vice-President Dick Cheney. Mr. Cheney left his position at the department of defense in 1995 to become CEO of the oil company supply giant Halliburton. Mr. Cheney had no prior oil industry or CEO experience. But he had plenty of government contracting experience, and insider knowledge.

Under Dick Cheney's tenure at Halliburton, government contracting business increased dramatically. Halliburton setup offshore subsidairy companies to get around U.S. sanctions to do business with Iran. Halliburton taxes dropped from $300 million in 1995 to an $85 million REBATE in 1999.

Mr. Cheney then became vice-president after the 2000 election. He immediately setup an energy task force called the National Energy Strategy Board, with undisclosed members from the corporate energy sector. Soon after, we all started to experience the results of Dick Cheney's energy program for America.

In March 2002, the New York Times revealed the names of 22 oil and gas executives who met in secret in February 2001 on Mr. Cheney's energy board. Nineteen of the oil companies were among the top 25 financial contributors to the Republican Party in the 2000 elections.

Under our supposed system of checks and balances, the comptroller general of the Government Accounting Office (GAO), filed a lawsuit against Vice-President Cheney for refusing to reveal the names of all industry executives in his secretive energy board, and what they discussed. The GAO dropped it's lawsuit after congress threatened to cut the GAO's budget. Once again, the media only reported these facts. They should have demanded accountability. Instead they simply moved on to the next human interest story.

What don't the politicians in Washington want the taxpayers to know? Why doesn't the newsmedia dig for the truth and demand justice and accountability? As the old political saying goes, all you have to do is follow the money.

## HALLIBURTON CONTRACTS

In March 2003, the Army awarded Halliburton and it's subsidairies, Kellogg, Brown & Root (KBR), and European Marine Contractors (EMC), a no-bid contract to rebuild Iraq's oil industry infastructure.

Under a no-bid contract there is no competitive bidding process. The government awards a contract to a "preferred" company without determing if others can do the work at a lower cost.

It should be noted that many of Halliburton's contracts have been awarded after a bidding process. However, they have still come under scrutiny. Senator Frank Lautenberg of New Jersey stated that all Halliburton and it's subsidairy companies contracts should be terminated. He also said;

"Halliburton's record of overcharging, bribery, and accounting fraud recites like a textbook example of corporate irresponsibility. Yet Halliburton has virtually monopolized the contracts in Iraq and has collected 9 billion through it's subsidairies."

Halliburton also received contracts to cleanup the hurricane torn gulf states region. Entire books have already been written on the Dick Cheney - Halliburton connection in this administration. Unfortunately the mainstream media has yet to embrace the facts and give credence to a full exposure and accountability campaign via relentless investigative reporting.

Some suggestions for controlling revolving door appointments.

* Revolk the special priviledges granted to former members of congress while they are lobbying government officials or working for industries that do.

* Place a 2 year ban after leaving political office before a person can own or work as a consultant to a lobbying organization.

The government watchdog organization The Revolving Door Group states on their website that 43% of the members of congress who left office since 1998 have become high paid lobbyists to the federal government.

## K STREET

K street in Washington, D.C. is to political lobbying what New York's Wallstreet is to stock trading. If a company, or industry, is trying to influence elected officials and department heads, then most likely it has offices on K street.

Once a Democratic Party stronghold of labor union representatives and manufacturers, K street lobbyists now run Washington politics like Wallstreet stock traders control the stockmarket. Buying favorable political policy with expensive trips and millions of dollars in campaign donations.

The elections of 1994 brought the Republican Party into control of congress. The Republican leaders brought with them a new party agenda they called the K Street Project.

The K Street Project was an attempt to take control of the K street lobbying firms from the Democrats. The plan to reach this goal was to get as many former Republican congress members, and or staffers, hired by the K street firms. In many cases former Republican office holders simply started their own lobbying firm.

A Republican on K street was practically unheard of prior to 1994. Now they out number the Democrats by a 2 to 1 margin.

## SUE, SUE, SUE; HOW WE GOT THIS WAY

There are many things you can associate with American society, and one of them is the threat of being sued. Attorney's claim it helps make America a better place. Large corporations say it is one of the reasons they re-locate into other countries. Eitherway when a person spills hot coffee on themself and can sue the person or business that delivered the coffee, things have gone too far. How did we get this way?

One of the biggest lobbying groups is the Trial Lawyers Association. And when nearly 80% of elected lawmakers on the state and federal level are attorneys, what other results could you expect but for them to take care of their own.

Our elected lawmakers have done a good job of passing laws and conduct rules that protect themselves if they get caught in a scandal or wrongdoing. That is not to say they won't go to jail if someone brings enough evidence against them. But keep in mind this real life example.

In December 2005, Congressman Randy Cunningham of California admitted to taking over 2 million dollars in bribes plus other compensation to help a contracting company get defense contracts with the federal government over a period of ten years. Representative Cunningham was a 16 year member of the U.S. House and a key member of the House Armed Services Committee. He was given an 8 year sentence, although unlikely he will serve all of it. He will still get to keep his federal pension estimated to be nearly $50,000 a year, even though he sold out his vote.

All Americans should be outraged over these situations, which most often go untold. It is bad enough he sold out his vote and congressional influence, but as a member of congress his taxpayer funded salary would be roughly $175,000 a year. And for his actions he gets to keep his taxpayer funded pension that is more than most Americans make in a year.

Although Representative Cunningham is not an attorney he surely has benefited from the rules they have made for themselves as congressional members. But that is what elected officials get to keep when caught in theft and bribery scandals. And it all comes out of the taxpayer's pocket. Shouldn't there be a law against that?

## CLASS ACTION

In mid-December 2005, a class action lawsuit was filed on behalf of the citizens of Alaska against the oil and gas giants Exxon-Mobil and BP. The lawsuit accuses the oil companies of purposely holding natural gas supplies in storage in Alaska in order to drive up prices by claiming a shortage of supply. This claiming of short supply obviously benefits all companies that make a living from selling natural gas, as well as commodity traders who ran up the price on the market.

The Alaskan citizen gets a government royalty check based on the amount of oil that moves through the Alaskan pipeline, as well as the amount of natural gas that leaves the state. This lawsuit is long overdue and hopefully will be played out in open court. Although that is highly unlikely. It will most likely be settled out of court with the oil companies not admitting any guilt, but paying the citizens of Alaska a small amount toward their royalty income, and the huge legal fees for all attorneys involved.

This kind of lawsuit, if pushed properly, can make the difference in exposing the oil and gas company efforts to drive up energy prices. It will take citizen's complaints and the right law firms getting involved on behalf of the people, and of course their own back pockets. Just as with the huge lawsuits against big tobacco companies in the late 1990's.

Until then, you will be paying more and more at the pump, and to heat your home. All conveniently blamed by big oil companies and commodity traders on the next big event. Not the true supply they keep manipulating.

## WHAT ABOUT THE TAXPAYER'S NEEDS

It seems most have forgotten that the founding fathers of this great nation had a concept of government for the people by the people. The needs of the average citizen of this country have been sold off long ago to the lobbyists for the sake of special interest money. All for the advancement of the Democratic and Republican Parties' agenda, which are not in  the citizen taxpayers best interest. They are at the taxpayer expense.

Whenever a politician gets caught in a scandal or wrongdoing, the political party leaders, and even the president, always have the same comment, once the person steps down from office. They commend them for stepping down for the "good of the party". They never say anything about for the good of the country, or taxpayers. Simply because the political party comes first and the special interest money a quiet second. Because it is what finances the party and the politicians. The taxpayers best interest are never met because of the manipulation of government laws and policy that special interest money influences.

Issues such as energy, medical care, and perscription drugs, are of interest to all Americans. A simple policy should be easy to achieve. However, the insurance companies, medical groups, and most notably the pharmaceutical industry have enough lobbyists on the state and federal level to get their people appointed to government energy boards. They ensure that their best interests are maintained by government policy, or lack thereof, by the very politicians the public elects and pays a large salary to, to protect the best interest of the taxpayer via government policy.

Some have suggested in the past that business contributions to politics should be greatly reduced or even eliminated. That the large energy companies influence in Washington makes it impossible to have a sound energy policy and affordable energy to consumers.

These are all things that the taxpayer not only needs, but deserves. Consumer spending should not be allowed to be controlled by the manipulation of energy prices by the large oil and gas companies and billionare commodity traders.

In my opinion, the political parties in this country are the backbone of all the problems. They serve no purpose for the citizen. They are simply industries unto themselves that have their employees and politicians in place to push their own agenda, which rarely, if ever, mirrors that of the American taxpayer.

What the taxpayer needs is truly independent candidates to elect to office that put the citizens best interest first. Only then can there be true governmental changes that can make American politics something for all to be proud of again. Image politicians honestly arguing over the best economic interest of the people for once. Not the best interest of their political party disguised as the best interest of the American citizen.

# CHAPTER 4

---

# WALLSTREET, BIG OIL,
# AND THE FIX IS IN

Every decade has it's political and business scandals. Junk bonds in the 1980's, dot coms and corporate accounting scandals of the late 1990's, just to name a few. The 1990's also gave rise to the start of the scandals of this decade. In my opinion, the 2000-2010 decade will be one of the most scandal laden decades in American history. Provided something is done about the current oil scandal.

Many of the indiscretions that occured in the 1990's have been brought to some level of justice after 2000. Big corporate entities such as Worldcom, Tyco, Arthur Anderson, and Enron, just to name a few of the notorious.

The same entities benefit from all of these misdoings year after year. Their corporate officials never have to admit wrongdoing or face jail time for their assistance in these matters. Their companies make enough political donations each year to both political parties to ensure themselves of protection and favoritism. (2005 Wallstreet contributions to political parties totaled $26 million). I am refering to the Wallstreet investment banking and brokerage firms who only pay a fine when caught or sued in federal court.

These same firms are also making billions on the current energy scandal. Wallstreet investment and commodity firms offer up their industry analysist to the newsmedia to spin their version of events they want to blame rising oil and gas prices on. The real truth will be explained in this chapter.

## WALLSTREET TRADERS

This political era of letting oil and gas companies manipulate the oil refining process to create huge profits has not gone unnoticed by Wallstreet traders. They jumped on the bandwagon and are equally to blame for high energy prices.

The speculative buying frenzy at the commodity trading firms has oil futures to the $60 a barrel plus range for similar reasons stock traders had investors paying $150 per share for overhyped dotcoms in the late 1990's.

Hedge funds, pension funds, private bankers, large capital investors, and even some university endowment funds are all trading in oil and gas futures now. The media is caught up in this current frenzy as well. Even the business news seldom reports the current cash price of a barrel of oil. They always report the nearest futures month price, which is usually 6 to 10 dollars a barrel higher than the cash price. And the CEO's of the commodity trading firms are making billions for their companies at the helpless consumers' expense.

Commodity trading is a paper transaction, either buying or selling of a listed commodity. Oil is the most actively traded of the commodities on the New York Mercantile Exchange. Simply put, it is a bet. Speculation that the price of the commodity will be either higher or lower in a future month of trading. A buy contract is speculating that the price will increase, while a sell contract is speculation that prices will be lower in the future.

Like any product or commodity, if there are more buyers than sellers then futures price will increase. If the are more sellers in the market then futures prices will decline. In either instance the current or cash price will move up or down depending on whether there are more buyers or sellers affecting it's movement. The cash price and nearest futures month price always move towards each other as the future month gets closer.

It is relatively easy to see how commodity futures traders can drive the current price of a commodity up or down. They trade with only slips of paper, without the need to see or possess the actual commodity.

It is obvious that if energy commodities, which began being traded in the 1980's, are going to be continued to be traded that the government needs more oversight and full regulation of the transactions. Crude oil, home heating oil, unleaded gasoline, oil distillates (diesel and kerosene), and natural gas, are all heavily traded in the futures market. It is important to keep in mind that these products are connected by the same source only. They are traded as independent commodities. Crude oil futures prices could fall, thus pushing down the cash price of oil. But unleaded gasoline, although an oil product, could remain high. Especially if speculative investors kept buying futures in unleaded gasoline. It would still remain high at the pump.

All commodity trading firms are regulated by the federal government. However, in recent years the over-the-counter (OTC) markets have emerged as a safe haven for energy commodity traders, because of the lack of government regulation in these markets.

The OTC markets have grown to over $248 trillion. It has allowed the large Wallstreet investment firms and hedge funds to go unchecked by the government regulators. Thus allowing them to easily engage in price manipulation strategies.

Another even more complicated strategy used by Wallstreet firms in the OTC market is the Derivative. A Derivative by definition is a financial security whose value is derived from another underlying financial security. Options and futures contracts are examples. Generally they are used to speculate on the movement of a commodity, but are also used to hedge against financial downside risk in the market. Simply put, a Derivative is a bet on a bet, or a bet against a bet.

Derivatives are extremely complex and their market actions are not fully transparent or even regulated by the SEC. This makes it a breeding ground for price manipulation. In the past Derivatives have been purposely mis-priced to conceal loses and make large profits by fraud.

The highly respected stock investor Warren Buffett said of Derivatives, "We view these as time bombs both for the parties that deal in them and the economic system. In our view, derivatives are financial weapons of mass destruction carrying dangers that while now latent, are potentially lethal."

A congressional investigation concluded that crude oil prices are effected by traders on the New York Merchantile Exchange as well as the OTC markets. They stated that the lack of information has made it impossible to determine if commodity traders are manipulating prices. Even after this conclusion, the Senate voted down a measure to regulate these markets in June of 2003. Once again, your elected officials paid with your tax dollars working hard for the special interest money they take, and against the American consumers best interest.

Some have argued against trying to get strict regulation of the OTC commodity markets. They rightfully claim that with all the special interest money politicians take from Wallstreet, it would be unrealistic to get laws passed anytime soon. Eventually the oil commodity bubble will burst just as the dotcoms did, and just as the stockmarket runup did once the accounting scandals could no longer cover up the truth.

The question that needs to be dealt with is why should Americans have to struggle to pay their heating bills and at the pump. We pay our legistators well to look after our interests. But instead they have sold out our best interests for large campaign contributions in exchange for letting large industries like oil companies and Wallstreet commodity traders make billions of dollars.

It is obvious it would be in the American consumers' best financial interest to have full regulation of all commodity markets. A more sensible solution would be to ban the trading of all energy commodities. Yet the speculative trading of oil and other energy commodities is freely allowed in order to let multi-billion dollar investment firms make even more money. All at the expense of the American citizen. This only serves to weaken the economy in both the short and long term. It forces too many consumer dollars to be spent on energy, thus further affecting consumer purchasing power. The energy sector is supposed to be a regulated area of the economy. Regulated for the purpose of a sound and stable national economy, and for national security interests.

Oil and gas prices are no longer dictated by supply and demand for product. They are now driven upward by speculative commodity markets, and the oil companies, that use even the most minor of events in any oil producing country to justify a price spike. Even in a worldwide market, if you take the oil commodity trading out of the price equation in regards to world events, oil and gas prices would not go up unless supply was actually affected. With commodity traders leading the markets, prices go up on only the perceived threat of a brief disruption. Making us all pay more for something that is not in actual short supply. With these results, common sense should tell us to elect people who are not associated with political parties that take large contributions from special interest groups.

## INVESTMENT BANKERS

The average American probably does not realize how much money the Wallstreet investment banks make off of nearly every financial scandal in this country. They usually have a hand in it, but their true involvement practically goes unreported. The recent corporate accounting scandal that took down many large corporations and their CEO's also went through the doors of most all the major Wallstreet investment firms.

In February 2003, Merrill Lynch paid an $80 million dollar civil fine for it's roll in the Enron scandal. Later that same year, J.P. Morgan-Chase Bank agreed to pay $162 million for criminal and regulatory misdoings in the Enron energy scandal.

In June 2005, Citigroup paid $2 billion for it's assistance in the Enron scam, but admitted no guilt. Also in June of 2005, J.P. Morgan-Chase agreed to pay $2.2 billion to Enron investors. J.P. Morgan and Citigroup each paid over $2 billion to investors for their banks involvement with the Worldcom accounting and stock scandals.

By August 2005 ten investment banking firms had paid large fines or reimbursements for their involvement in the Enron case. J.P. Morgan-Chase also agreed to pay $1 billion to the Enron bankruptcy estate. These are just a few examples.

No person working for any of the banks went to jail or even to trial. Yet what the firms paid in every case is a huge sum. But not large enough to deter their involvement or actions to assist in one case after another, year after year. Which tells us the fees and profits they make are certainly much greater than the fines imposed under current law.

The firms that specialize in commodity futures trading are not in it alone. All major investment banks have a hand in commodity trading as well. There are large fees and huge speculative profits for them to make in futures. Especially in the OTC markets.

Does all this mean they are corrupt and should be put out of business? No. But the nature of big business and speculative trading always seems to lend itself to some level of wrong doing in the pursuit of large profits. Throw in some politics and you have a scandal. It all comes at the expense of the taxpayer and investor. Costing millions of dollars to investigate and prosecute these cases.

What is needed is full accountability for all parties involved in any financial wrong doing. The people at the large investment banks are the best and brightest in their field of employment. They should not be allowed to simply walk away and let the bank pay a large fine for their indiscretions. The fines should also be much greater than any profit they may have reaped. Remember, the driver of the get-away car is just as guilty as the guy who goes into the store and robs it.

## OIL COMPANIES MANIPULATE PRICES BY REDUCED REFINING

As mentioned in the previous chapter, Vice-President Dick Cheney's formation of the National Energy Strategy Board was nothing more than a call to Washington for the big oil and gas executives. It gave the large energy companies the green light to do what they had merged in the late 1990's to do. Manipulate the refining of oil to justify higher gas and heating oil prices.

The National Energy Strategy meeting probably amounted to nothing more than a simple meeting, at taxpayer expense. It obviously let the oil and gas companies know the new administration was not going to give credence to, or act upon the regular reports that the federal agencies that monitor the oil and gas supplies would be filing.

As consumers we have all seen oil and gas prices briefly fluctuate slightly in response to events, or extreme weather. But only since Dick Cheney's National Energy Strategy meeting do such events have such a profound price effect to the consumer.

When asked by the media who attended his National Energy Strategy meeting, and the details of what they discussed, Vice-President Cheney involked executive priviledge. He claimed so on the basis of national energy security. But this meeting was held prior to September 11, 2001. Regardless of which event had happened first, the purpose of elected officials dealing with energy strategy is to secure the nations energy needs. It would involve measures that guard against price manipulation tactics, not encouraging it. It would involve securing a more stable supply in order to reduce price fluctuations and guard against event driven price runups. Thus keeping consumer energy prices low and reliable. It is obvious protecting American consumers from price fixing was not Mr. Cheney's goal with his energy board.

If elected officials were really concerned about controlling energy prices they would not allow energy to be traded as commodities on the futures market. Where the price of oil and gas can be easily manipulated.

Simply put in regards to Vice-President Cheney's National Energy Strategy Board; There is no executive priviledge to conspire with energy companies to shakedown the American consumer. Yet it has been allowed to continue. If the Democrats weren't taking large contributions from oil companies as well, then they would have been screaming for impeachment hearings long ago.

Consider these facts verses the hype that is reported in the news media. The U.S. Government has an agency with it's own website that monitors and reports oil and gas supply and demand in this country, and in the world. The U.S. Energy Information Administration (EIA) reported that world oil demand increased 3.32% from 2003 to 2004. It's figures also show that world oil supply increased during the same period by over 4.4%. Equating to nearly a 1 million barrel per day surplus of oil worldwide. Their reports also show total U.S. petroleum supply, not counting the strategic reserve, which has also increased, was almost 25% higher in May of 2005 than in 2003. Yet oil companies and Wallstreet commodity firms have their paid analysts tell the media just the opposite. That there is a worldwide shortage of oil, and that U.S. refineries can not keep up with demand.

The Federal Trade Commission reported to congress in the fall of 2005. They stated that oil refineries in the U.S. were operating at well under 80% capacity in May. Well before hurricane Katrina. This reduced capacity output also mirrored their reports from the previous two years. They went on to conclude that high gas prices were the result of decreased refining in recent years, not lack of oil supply, which in fact had increased, as stated in the EIA's reports.

There are several government watchdog websites, as well as consumer rights groups, that have reported congress's lack of action over EIA and FTC reports. You can log on to their sites and read the various government agency reports. Many of these sites also have internal memorandums that date back to the mid 1990's from most all the major oil companies. The memo's all refer to the need for the oil and gas industry to reduce nationwide refinery output in order to increase profits.

Since the mid 1990's several small oil refineries have been bought and shut down by one of the five major oil conglomerates. BP, Exxon-Mobil, Shell, Chevron-Texaco, and Valero Energy, are the top five oil and gas companies that were once fourteen different companies prior to 1995. Everyone else in the industry plays along by their rules, or gets squeezed out of the market then bought out.

## A CORPORATE CARTEL AT WORK

Although cartels are illegal in the U.S., the five major oil companies have been able to manipulate the market just the same. Any organized attempt to control and manipulate a product or market in a free market economy is prohibited on the basis of consumer protection and fair business practice. However, that's what millions of dollars in political donations has allowed the oil giants to do.

In comparison, the OPEC cartel operates outside the U.S., and is a group of independent nations that have ageed to control their oil output in order to manipulate prices. The top five oil companies in this country are huge corporations that merged for a common purpose. To control oil refining in this country, and the flow of natural gas, in order to drive up prices and huge profits for themselves. All the signs of a cartel at work. Manipulating the product or it's movement for greater profits.

So why is this U.S. 'Corporate Oil Cartel' so much more effective than OPEC at manipulating the oil market?

OPEC nations have plenty of competition from Non-OPEC nations in the worldwide market. Oil is their country's only major source of income. So they can never cut production to extremely low levels without hurting themselves.

The U.S. 'oil cartel companies' have merged in order to control all refining in the U.S. They have bought off both corrupt political parties in order to make sure their illegal actions go unpunished. They figured out in the early 1990's that the real financial control of gas and heating oil was not determined in the supply of crude oil as much as it is in the refining of oil into usable products. In otherwords, you could have crude oil stockpiles so large in this country that we won't have a place to store it, and still have high gas prices if you allow big oil companies to shut down refineries and continue to operate the ones at industry discretion at a fraction of their capacity. Which is what has been done.

## HOW DO WE STOP THIS TRAIN?

Energy conservation by all consumers is always a good measure to follow. However, this alone will not end the current energy price runup. The root of the energy crisis in this country is not limited to the manipulation of oil refining by oil companies. Nor is it the responsibility of the Wallstreet commodity traders alone. They are equally to blame.

The root of the energy problem, like most issues in this country, is at the behest of the politicians. The answer to the problem is within the political structure as well. The political parties have become so corrupt that their politicians can not be trusted to perform even the most basic services for the public they represent without undo corporate influence. All citizens must standup for their interests in order to bring about political and economic change.

In December 2005, a class-action lawsuit was filed on behalf of the citizens of Alaska against BP and Exxon-Mobil for holding natural gas. The suit was the first, and hopefully more will come. This lawsuit will not be a saving grace for us all. It will surely reap great rewards for the attorneys. Hopefully, it will serve as a shot across the bow for a sign of things to come.

With our political system broken and the newsmedia unwilling to investigate and pursue the issues, class-action lawsuits are our best hope of bringing about an end to refining manipulation, and political corruption. It is truly a shame that some egregious lawfirm will prosper doing what our tax dollars pay our elected officials to do in the first place. Maybe there is a group of attorneys with enough moxie to take on the political parties as well. But without a big payday for them at the end, it is highly unlikely.

It will be up to the voters to force political change by electing new faces. They must voice their demands that political lobbying must end, and all involved be held to a higher level of accountability than what is current. They must also demand a ban be placed on energy commodity trading to end the speculative price runup in oil, gasoline, and natural gas.

Not being named in a political lobbying scandal is not a reason to re-elect someone. If they have not previously stood up with a bill to end corporate lobbying and large campaign contributions then they are equally culpable for doing nothing while the American consumer is fleeced.

Without true independent candidates to vote for, nothing short of banning political parties in this country will bring about a true change for the betterment of the people and government policy. A government should always be about what is best for the citizens. Not using what the citizens need as a means for large financial gains for the political parties and the corporations that fund them.

## CHAPTER 5

---

# U.S. ENERGY POLICY

## WHAT WE NEED AND WHY

What America needs to cure it's constant energy concerns is a sound national energy policy. A policy that serves the best interest and needs of the consumers with a stable energy supply. Not one that is dictated to the people through government by the big oil, gas, and electric company lobbyists. Our current energy program is at best bad policy, but more appropiately a political scam for the financial benefit of the energy industry.

America's energy needs can be met from North and South American supplies. They should be primarily met from North American sources. With all the bio-fuel technology at our disposal there is no need for the U.S. to depend on the middle-east for oil. Yet politics and special interest money are the stumbling blocks to this goal that should be standard policy.

If the government had put the money we have spent on behalf of the big oil companies in the middle-east over the past twenty-five years into bio-fuels and other technology, America could possibly be a net exporter of energy. Certainly we would be able to derive all our own energy needs from our own hemispere.

U.S. energy policy should be forward thinking and self-reliant motivated. We should use advanced technology applied towards products to create fuel and energy from domestic sources. It is imperative that we overcome and move past the corrupt indiscretions of the past and current energy policy. It is infact a political policy and not a true energy policy. It relies too much on what the corporate oil cartel want, not what is in the best financial interest of this country and it's consumers.

This country's energy policy should not simply state that it has a goal of reducing our dependence on foreign oil and stabilizing energy costs. It must have basic common sense measures to meet these goals.

For all the reasons mentioned in previous chapters, the U.S. Congress must ban the trading of all energy commodities if it is really interested in controlling consumer level energy prices. It also needs to breakup the recent mergers of the big oil companies.

## MPG STANDARDS

Federal highway mileage standards for all cars, trucks, and S.U.V.'s must be increased for the next model year. We have the proven technology for sale in the automotive aftermarket for a few hundred dollars per vehicle. It needs to be done at the factory on all vehicles immediately. There is no reason to give industry the typical three to five years to comply on something that can be done with existing equipment in a matter of minutes.

## BIO-FUELS

America being the leader in agricultural production should have bio-fuels as an important component to our energy supply. The oil industry has fought against this competition for years. Saying everything from it is too expensive to cars won't run on it. The broader question is how much money does a barrel of oil really cost, when you add in all the government support and foreign policy matters that cost trillions of dollars. All in the name of oil profits for the big oil companies at the expense of the taxpayers.

To further the immediate use of bio-fuels, the government should change the old federal EPA laws for metropolitian areas throughout the country. The laws now require some fifty different blends of gasoline nationwide during the summer months to meet EPA air standards for clean air. The use of, or mixture with, bio-based fuels such as, but not limited to, ethanol or soydiesel could greatly reduce the number of fuel blends. With different blends possibly being required for the northeast region, west coast, and the large cities of the southeast and midwest states. Reducing the number of blends would speed up the refining and distribution process.

## KING COAL

America is a vast country with vast natural resources. In this current era of out of control oil and gas prices we tend to forget about coal. As discussed earlier in this book, coal has been used to make fuels as well as natural gas. These are important uses for a resource that is still an important part of the energy equation. If this country really had an energy supply shortage, coal gasification and liquification would be getting plenty of promotion. Still their use should be advanced and become part of our energy supply.

## ALTERNATIVE ENERGY AND NET METERING

With this nations extensive river system and countless dams thereon, it would be a wise investment in government and university research dollars to develop efficient means to install hydro-electric generation upon existing dams. Other areas of note in the use of alternative fuels are anaerobic digesters. Digesters take plant and organic matter and turn them into natural gas. Another option already in use in Eastern Kentucky involves taking the methane gas from landfills and using it on site to generate electricity. A landfill based methane gas facility costs about $2 million, and has a life span of about 20 years.

With the strain being put upon this nations electrical grid system the expanded use of alternative energy sources such as geo-thermal, wind, solar, and hydro, should always be an option to expand upon. To make these systems more affordable and accessible to both homeowners and small businesses, a federal mandate is needed to enact true net-metering. Net-metering allows a person or business to generate their own power via solar, wind , or hydro, and sell the excess power they may generate to the power company. When producing more power than they use the excess power makes the electrical meter turn backwards. Thereby giving the customer a reduced electrical bill, or possibly a credit on their monthly bill.

Many states have pushed for net-metering laws. But the rules are still dictated by the power companies who help write the laws and thus make it difficult and freely accessible for just about any person to use. California is one of the few states I know of that has manditory net-metering laws. California also leads the nation in solar power development as well, although special interest groups lobby against it's use.

Because of their large purchase prices, larger federal and state tax credits should be given to individuals and businesses who instal geo-thermal, solar, or wind power units. Local tax boards should also give an annual tax credit of at least 10% to a home or business being powered by one of these sources.

## WHY CAN'T WE GET SOMETHING SO SIMPLE?

Before I address the question, let's recap our energy needs.

*Ban the trading of all energy commodities.

*Decrease use of oil from mid-east.

*Increase gas mileage on all vehicles.

*Increase use of bio-fuels.

*Increase use of coal and other domestic products to make natural gas and fuels.

*Make better use of alternative energy sources, and larger rebates for their use.

*True electrical net-metering for all consumers.

*Remove/ban oil, gas, coal, and electrical industry representatives from all government boards.

The energy needs of this country are simple, given our natural resources and level of technology. It is the control, use, and profit, from these resources that compels entire industries to influence government policy through large financial donations to both political parties. This is what takes simple needs and turns them into complex issues.

It is because of these large campaign contributions that mayors, govenors, and presidents, appoint directly or indirectly, via government agencies, industry representatives to energy boards. It is a good idea to consult with several different industry representatives to better understand energy use and needs. However, it is an absolute lack of responsible leadership to appoint insiders to an energy board to frame a policy that will directly effect their industry and therefor their personal income. Common sense dictates that nothing other than a policy that protects the industries best financial interests could come from such a board. The industry representatives are there to do nothing more than protect their respective industries share of the energy pie. They work together to make sure alternative resources such as bio-fuels are kept on the back burner, or limited by the very energy program they help construct.

There is a short answer to the energy needs question. Why can't we get something so simple? When you combine politics with industry influence via lobbyists money, it always equals corruption, and therefor bad policy and government for citizens.

As long as there are politicians that accept bribes there will be corruption. As long as there are political parties that make laws that allow them to collect large sums of money from special interest organizations, then government policy and the citizens' best interests will continue to be sold off to the highest bidder.

## SUPPLY AND DEMAND DRIVEN IN A FREE MARKET ECONOMY

Everytime someone in the newsmedia interviews an oil and gas expert, or a Wallstreet commodity trader, they tell us we will never see gas under $1.50/gallon or oil under $40 a barrel again. They all state that it is a different era now. They say we are now in a global economy and must compete with Europe and Asia for the world's oil supply.

Global competition is nothing new. Just hype to justify the oil industry traders current financial position in the market. The fact is we have always been in competion with Europe and Japan, and other parts of Asia for oil. True, Chinese and India are gaining U.S. jobs. But they are gaining them because of the extremely low pay.

China and India buy all the oil they do not produce themselves from nations that have plenty of oil reserves, but have not sold oil to the U.S. for a quarter century. Nations such as Iran, Sudan, and previously Iraq, would rather go broke before selling oil to the U.S. Even with Iraq's oil fields under U.S. control, the oil is still sold to the same nations as before. This excludes the U.S. The difference now is that U.S. companies are making profits from the oil instead of Saddam Hussein.

In January 2006 a Wallstreet oil trader estimated that the uncertainty over Iran's nuclear program was adding $10 to $15 a barrel to the U.S. market price, via speculative oil commodity trading.

Action in Iran could briefly play a roll in world oil supply, maybe. But it has little to do with actual or current supply and demand. That is why it is known as speculative trading. Commodity trading needs an event to make prices move quickly in order for traders to make big profits. Simple supply and demand is too stagnant to create large profit opportunities. Commodity trading has obviously had a huge impact on the price of oil and gas to consumers. This is why energy commodity trading should be banned.

In January 2004 oil was considered being kept high at $28 a barrel by OPEC. U.S. E.I.A. data going back to 2002 shows world oil and natural gas output outpacing the rise in demand. So where is the justification for the leap in oil prices to $60 plus per barrel based on supply and demand? There is none. Since 2002 we have simply felt the full effect of Vice-President Dick Cheney's secretive National Energy Strategy Board. When politicians allow oil and gas companies to manipulate supply, it really does not matter what demand actually is.

Looking at supply and actual consumption of oil yearly since 2002, it is hard to justify oil over $20 to $25 a barrel with out price manipulation. With commodity traders gone wild and oil companies financially influencing politicians, you get  $50 to $70 a barrel oil. OPEC, which used to get the blame, is now and afterthought. True supply and demand have been thrown out the window, and the energy market is now event driven.

Oil and natural gas will go higher without corrective intervention. It will take newly elected officials who are willing to ban energy trading. An independent  attorney general is needed to fully investigate and prosecute all involved in this corporate and government oil and gas supply scandal.

## ENERGY SHORTAGE NONSENSE

In December 2005 the U.S. E.I.A. reports showed natural gas and oil supplies in this country were at five year highs. Some continue to argue that high prices are simply a reflection of supply and demand, even when the independent facts clearly show otherwise.

The mild U.S. winter of 2005-06 normally would lead to falling fuel prices. However, with fuel industry and Wallstreet traders given a free hand in our current political environment, they simply turn to the next world event to continue to justify high prices. As we have all seen on the news, this equates to record profits for the oil and gas companies. All at your expense.

No longer do these markets trade on actual supply and demand figures. Energy commodity markets are now driven by large investor dollars on the basis of fear and greed. Fueled by any event they can use to drive up the market, and their profits. All this while the politicians simply look the otherway.

Oil and gas industry experts have made claims of the world running out of oil for nearly a century. Driven by the need for government support and tax breaks, as well as to bolster profits. These predictions have never come true. Common sense tells us that if this were true that we would have had a comprehensive energy plan decades ago.

The few so called shortages that we have experienced have all been man-made. Born of true greed and manipulation. Both being for monetary and political gain. As always, the consumer pays for the results of the devious actions of those benefiting from the profiteering. Just ask those who waited in long lines during the 1973 Arab oil embargo. The big oil companies got rich on the oil and gas they had stockpiled, while American's were rationed gas as if there were no other world source.

Research and technology have always made America a world leader in every aspect, including energy development. If allowed to blossom to it's full potential, alternative energy sources will keep us from ever running out of an energy supply. We simply need leaders who are not joined at the hip with corporate oil money, and are truly willing to end America's dependence on mid-east oil.

Oil will continue to play an important roll in our energy needs for decades to come. The alternative options have rarely received the investment dollars needed to be viable options until now. Still much more investment is needed. The U.S. Department of Energy has signed an agreement to build FutureGen. FutureGen is a prototype powerplant of the future. It will produce electricity via hydrogen with zero emissions. Peabody Coal is also building a coal plant in Illinois to turn coal refuse into natural gas. Both are steps in a positive direction to help alleviate our dependence on oil from politically unstable areas of the world.

Fortunately, we have never had a full blown nationwide energy shortage. What we constantly have is a shortage of honest leadership that has this country's, and it's citizens, best interests in mind when it sets forth government policy.

## CHAPTER 6

---

# PUTTING IT ALL TOGETHER

## CYCLICAL MARKETS

All energy resources have cyclical prices. They are higher for periods of months or years, and then they are lower once again. World events, political policy, and seasonal weather changes can and do influence supply and demand of energy, thus affecting price.

Coal provides over 50% of our energy needs, but it is the least affected by world events and seasonal changes. This stems from two reasons. It is not an actively traded commodity, and it is a domestically produced resource. Like oil and gas, it's price is cyclical, but subject only to the government's EPA policy regarding sulfer content and clean air standards, not to world or national events. Thus making it more stable in price.

The battle between industry and government policy is cyclical as well. The oil and gas industry wants to drill in the Artic National

Wildlife Reserve (ANWR), which is a government protected area. In decades past it has been areas of restriction such as other wildlife areas and the issue of offshore oil rigs that were in demand by the oil giants. In every instance the industry claims we are quickly running out of oil. They raise prices, along with campaign contributions, in order to get congress to change policy and allow them drilling rights.

Only in this current political era have we seen our executive branch form a secretive 'National Energy Strategy Board' with the oil and gas executives. Then claim executive priviledge when questioned about the actual strategy and purpose of the board by the press. They followed this with swift and heavy-handed political tactics to squash the brief government inquiry into abuse of executive power by the GAO.

Only one conclusion can be drawn from the situation of oil executives making large campaign contributions, then being asked to secretly form an 'energy board'. It should be no surprise that energy prices and oil industry profits have skyrocketed ever since.

## THE GANGSTER FACTOR

Most people believe that the 1990's were the first era of the average U.S. citizen being invested in the stockmarket. Not true. The 1920's were the decade that first saw the average citizen investing in stocks.

It was all the rage of the roaring twenties to be investing in the stockmarket. The average working man owned a home, a car, and had some stock investments. They understood little of what really drove the market, only that it continued to rise during this period of seemingly endless good times.

In 1927 a college economics professor wrote a newspaper article warning against the false rise and inevitable crash of the stockmarket. He based his statements on sound market fundamentals of trading that were no longer being displayed on Wallstreet. His article got him fired from his teaching position, and no other college would hire him. His outspoken comments, although based on economic and market fundamentals, made him an outcast. Some even called him unamerican.

In that same year of 1927, Chicago gangster Al Capone was on trial. As always he was acquitted of his mob crimes. When he left the courthouse reporters swarmed him, asking a bounty of questions. In a lighthearted jesture one reporter asked if he was investing in the market like everyone else. Al quickly replied, "Now there's a scam if there ever was one." He went on to say, "there is a handful of guys up there that control the market, and they ought to be arrested. But they won't because they payoff the politicians. Mark my word, when that scam blows up the market will crash and those guys that control it will make off like bandits and the common man will be lucky to keep the shirt on his back."

Al Capone and the economics professor were both right. Oddly, more people took note of what Al said than the professor. Two years later the market crashed in 1929, it was revealed that in fact seven men, all prominent investment bankers and wealthy investors, had formed a group and schemed to drive up the price of certain stocks. They bought and sold amongst themselves on the open market in order to drive up prices. Then they dumped the overpriced and overhyped stocks on the average investor.

Al Capone also told the reporter that the difference between him and the guys on Wallstreet was that what they were doing was going to hurt the whole country. "Tell the cops to arrest those crooks on Wallstreet and leave me alone," he was reported as saying as he walked to his car.

Trading on Wallstreet has always been ripe for manipulation by the big players. There is always a multitude of manipulative practices going on in order to forge large profits for those that hold the most cards. Few practices have the political backing to truly take on a life of their own and become obvious to all, such as the corporate accounting or current oil scandal. I bet Big Al would not be a bit surprised at the oil scandal or any of the past corporate or stock scandals since the mid 1990's. He would probably laugh and say, I told you so.

As the old saying goes, the more things change, the more they stay the same. Still the stockmarket remains a sound long-term investment. You have to be willing to accept the highs and lows in a cyclical market, regardless of the cause. All markets eventually return to the fundamentals of trading, no matter how big the scandal.

## KNOWING WHAT TO BELIEVE BY DISSEMINATING WHAT YOU HEAR

Someone once said the first casualty of war is the truth. The same can be said of most conflicts, political campaigns, and industry spin made by analysts.

In politics and business, analysts and insiders are always trying to spin, or sell you on, their version of events. All in the name of political power and monetary gains. When watching the news, business reports, or when reading the paper, you always have to ask yourself what are they trying to sell me on. And what will they, or the industry in question, gain from it.

In today's media, the news is driven more on thirty second sound bites than on true investigative reporting. You have to always consider the source in any news report. Was it simply reported from the politicians and or industry in question, or were the facts gathered via an independent investigation.

In recent years independent polls have shown that more Americans get their political news from the late night comics of Jay Leno, David

Letterman, and Conan O'Brien, than from the evening newscast. On one hand this is a sad comment on American's. On the other, there is a good reason for it.

The latenight talkshow hosts all use current events and bad politics to polk fun at during their opening monologue. The high price of gas directily linked to record oil company profits and corrupt politics are almost nightly staples. It is a matter of connecting the dots to show the corruptive link, then making a joke of it. Although it is no laughing matter.

In contrast the nightly news simply reports the record oil profits when they are released by the company. They briefly reported on Cheney's secretive energy board, but never followed up with good reporting that would have exposed the political wrongdoing and demanded justice until gotten. Instead they give more attention to human interest stories, because they take less effort to report.

The fact is, on a nightly basis, Leno, Letterman, O'Brien, and the others, do as good a job of informing the public than any of the major network newscasts. They keep the government and business scandals on the minds of viewers by continually making honest jokes about them, long after the major media outlets have moved on to the next in-depth human interest story or natural disaster.

When a news story breaks, it's frontpage news. What you read will always be mostly speculation and hype just to get your attention, as few facts will be known. A week later the same story will be a backpage article, a fraction of it's frontpage length. This is because the facts will be covered and they usually are not as interesting, nor do they take up as much space as the speculative hype.

Maybe the evening newscasts should give a couple of minutes each night and show their respective latenight talkshow host's coverage of political events. Better yet, why don't they, and the newspapers, just do more in-depth investigative reporting in order to keep politicians and big business honest. They need to follow political wrongdoing from beginning to end. Not just until the next story breaks a few days later.

Politicians and corporate executives used to fear the media because of exposure for dishonest practices. Now they use the media to tell what they want everyone to believe as fact. It is called spin. And the media generally reports it with little questioning or followup reports to set the record straight.

I heard a long time ago that a smart man knows to believe only half of what he hears, and an intelligent man knows which half to believe.

## WHEN WILL HIGH ENERGY PRICES END

In the spring of 2004 an executive with Exxon-Mobil said in an interview that current supply and demand of oil was not the cause of $50 a barrel oil. He put the blame on the energy commodity traders that continue to buy oil and gas futures at alarming rates, driving the market skyward.

In the wake of the September 11th attacks oil and gas prices soared. An oil and gas industry analysts stated in a t.v. business report that with true supply and demand forces at work, for every $1 dollar a barrel oil rises equates to two-and-a-half cents at the pump. This accounts for refining and transportation costs to the gas station.

The speculative commodity trading that is part of the problem will continue to keep fuel prices abnormally high until the market crashes. Remember markets are cyclical. And just like tech stocks that went out-of-sight in the late 1990's only to crash in 2000, eventually so will energy commodities. Although it will take electing new members to congress who will pass corrective legislation to ban energy commodity trading to do so.

Unfortunately we, nor our elected officals, seem to learn these lessons very well. They continue to repeat themselves. First in one market, then another. Always for a lack of oversight and regulation by the politicians we continue to elect to do just that.

In the wake of the Enron energy scandal, a few senators have introduced bills that would have fully regulated energy commodities. However, they were soundly defeated everytime. For elected officials to allow the unregulated trading of energy commodities to continue, especially in the wake of Enron, is a breach of responsibility.

It is ludicrous to allow speculative trading that can only hurt the consumer. Only when the citizen consumer gets involved and demands changes in politics by electing new representation will energy futures trading end. The high price of energy will not fall back in line with true demand until justice is demanded by the citizens for the oil and gas companies that have manipulated gas supplies for their own financial benefit. Many have already brought these issues to light. It will take a collective effort of citizens demanding justice before justice will be served.

## WHAT YOU CAN DO TO CHANGE THINGS

Polls in recent years have shown that most American's believe that politics are so corrupt that no matter who gets elected it really does not make a difference. Unfortunately this is pretty much the case, with the lock the two political parties have on elections in this country. However, this does not mean a person should not vote. It is good to limit those in office from staying there if they are not drafting legislation to change political policy for the betterment of the people.

Voting for a candidate is like hiring an employee. Elected officials are suppose to represent the people, although in the last two decades the two parties have sold government policy to the special interest organizations. Sometimes this even includes foreign governments. A good place to read all the facts on this matter is at whitehouseforsale.org, which is a public citizen web link.

It is not a good idea to keep an employee (i.e. elected official) around for very long. Especially when they are prone to giving themselves pay raises and better retirement and insurance benefits than their employer (the taxpayers) have. When the people's needs are continually being sold off to corporate lobbyists, then it is time for the political establishment in both parties to go, or simply both parties.

We need more representation on the behalf of the people. We can not stand any more political representation for the special interests of big oil, healthcare, and the pharmaceutical industries, at the people's expense.

In order to bring about change, citizen involvement must not end at the voting booth. All citizens must demand change from current elected politicians. Phone calls and emails to state and federal officials is the easiest way to let your thoughts and feelings be known.

You can not expect political parties or their politicians to police themselves. That is the equivalent of expecting toddlers to run a daycare. If the American people do not get involved and demand change and justice for current and past discretions, then we will continue to pay higher energy prices, and higher medical costs, while our elected officials live like kings at the expense of the taxpayer.

Demands must also be made of the national newsmedia for more in-depth investigations. Not just of the corporate lobbying scandals, but of the true reach of money in politics, as well as the people and politicians involved. Only then will change actually be instituted.

## FINAL THOUGHTS

The only positive influence that can be seen from high energy prices is that it forces all Americans to be more conservative in their energy use. The demand for more energy efficient cars and other products is on the rise. But we have seen this cycle before. Hopefully, alternative domestic sources for energy will get a lasting foothold in the American energy equation this time.

In past high cycles, oil miraculously drops in price to affordable levels when bio-fuels and other alternative energy sources get increased government funding for mass-production, or during some election years. Some would argue this is not downside manipulation by oil companies, but simply a reflection of consumer conservation measures decreasing demand. Still the same companies continue to lobby the political parties for greater profits at the expense of the American citizen. And the cycle is unfortunately repeated.

On occassion, when gas prices get high enough to make consumers angry, the oil industry will have one of their analysts make a stark comparison of how cheap U.S. gas prices are compared to Europe and Japan. What they do not tell you is that the tax on U.S. gasoline is figured in cents per gallon, while European countries and Japanese taxes on gasoline and diesel fuel are figured in dollars per gallon. Each U.S. state gas tax varies, but California and Hawaii have the highest tax at just over fifty cents each. There simply is no logic in comparing U.S. and European gasoline costs at the pump, unless you also report this major difference in gas and diesel tax and then deduct it from the equation.

Until the American voter decides to vote for true change at all levels of government, the large corporate interests will continue to control politics of both the Democratic and Republican Parties. Until the American people force change upon the two corrupt parties that make political policy in this country, then we only have ourselves to blame.

**Other books in print by J.C. McElroy include the fiction titles:**

"The Bluegrass Boys: A Story of Politics, Horse
Racing, and Marijuana Smuggling"

"Tres Novellas: The Dream Weavers"    (short stories)

## ABOUT THE AUTHOR

J.C. McElroy has a degree in Economics and Business Management from the University of Kentucky. He has written several books including both fiction and non-fiction titles.

Made in the USA